YOUR KNOWLEDGE HAS VALUE

Bibliographic information published by the German National Library:

The German National Library lists this publication in the National Bibliography; detailed bibliographic data are available on the Internet at http://dnb.dnb.de .

Imprint:

Copyright © 2017 GRIN Verlag
Print and binding: Books on Demand GmbH, Norderstedt Germany
ISBN: 9783346022066

This book at GRIN:

https://www.grin.com/document/500559

Nadiia Kudriashova

Reconstructing Society in the Post Slavery Caribbean

GRIN Verlag

GRIN - Your knowledge has value

Since its foundation in 1998, GRIN has specialized in publishing academic texts by students, college teachers and other academics as e-book and printed book. The website www.grin.com is an ideal platform for presenting term papers, final papers, scientific essays, dissertations and specialist books.

Visit us on the internet:

http://www.grin.com/

http://www.facebook.com/grincom

http://www.twitter.com/grin_com

Reconstructing Society in the Post Slavery Caribbean

The French Revolution brought the Caribbean slaves equality, fraternity and the long-awaited freedom. Negros and mulattos of Haiti demanded that the slogan of "liberty, equality and fraternity" was applied to Haiti as well as to France. To defend their demand, they conducted for decades one of the most severe and most bloody wars in the history of Western half-globe. Haitian Revolution, being one of the most important episodes of struggle of liberatory fight of modern times, found a wide response worldwide. It has helped to undermine the foundations of slavery system. The victory of the revolution in Haiti dealt a serious blow to the colonial system and contributed to the rise of the liberation movement in Central and South America. Independent Haitian government provided military assistance to the fighters for independence in Latin America. For the termination of the slave trade Spain paid UK about 400 thousand pounds. Although in Cuba slavery was finally abolished only in 1886.

The Haitian Revolution, which at critical moments in the struggle against the Latin American colonial oppression gave refuge and inspired many revolutionaries of the continent, has also had a special and powerful influence on the formation of the Cuban nation. Latin American countries after liberation from Spain became formally independent republics. The stage of formation and development of politically independent Latin American countries and respectively the Latin American nations began.

The leader of the Haitian Revolution, Toussaint L'Ouverture, "taking into account the importance of his services rendered to the colony", as "a sign of the unlimited trust" for him on the side of the people was declared ruler of San Domingo with the right to appoint a successor. The governor of the colony was both commander in chief. He was responsible for internal and external security of the country, has the authority to communicate directly with the mother country of all questions relating to the interests of the colony. To him the legislative initiative belonged, he signed laws, appointed all civil servants and the commanders of military units, was the chief censor. Legislative body of 1801 Constitution was the central meeting, which included representatives from each department (Fick, 2007).

Back in October 1800 a regulation was issued that prohibited farmers to abandon their plantations and pass to others. The Constitution of 1801 maintained the prohibition. In addition, sale of state land plots smaller than 50 acres was prohibited; with acquiring land, the owner had to guarantee its economic use. In other words, compulsory managing og plantations was assumed.

The relationship between employees and management of plantations (whether appointed by the governor, or former owners) was defined as the relathionships of active and strong family, where the father is necessarily the owner of the land or his representative ... Every employee – is the member of the family income and equity holder. The practice of life was far from idyllic relations stipulated by the constitution. The remaining planters used all possible attachment of workers to the ground for ruthless exploitation. New plantation managers often proved no better conditions than the former owners. This was causing dissatisfaction of former slaves. However, Toussaint Louverture introduced and strengthened the plantation system (Fick, 2007).

The resulting state was a state of transition. It preserved monocracy inherited from the war had just ended and the resulting of still not completely eliminated threat posed by external and internal enemies. In it, on the one hand, freed slaves who have received land, tended to the creation of the peasant economy, and on the other – the interests to strengthen the freedom won by them caused the need to maintain the plantation economy. The state could not defend the freedom of the Negro without placing a large and well-equipped army, control apparatus, fleet, means for the restoration of the war-torn roads, bridges and public buildings, for organizing mail, etc. Emerging farm in any way could not satisfy all these needs. At first it was supposed to have mainly natural character and was not able to provide even a meager diet of the population of the colony. In addition, the skills of slaves in the conduct of the peasant farms were clearly insufficient and confined mainly to the ability to cultivate small gardens. Conservation and restoration of sugar cane plantations and coffee allowed to solve the problem, challengingat the time the state of former slaves. Trade in the products of the traditional economy made it possible to acquire the necessary funds to purchase food (they did not have enough in the war-ravaged country), maintenance of the armed forces of the state apparatus, etc. (Fick, 2013).

So, in the name of the final consolidation of freedom it was necessary to support the plantation, in some extent limiting the freedom of action of the former slaves. The state exercised daily control over the plantation owners and managers of plantations through its commissioners. Corporal punishment was strictly prohibited by law. Severely were punished abuse, theft and mistreatment of workers. Toussaint personally go into all the details, keenly watched to free Negro was not gradually turned into a slave again. To the residence of Toussaint anyone could come and bring a complaint against anyone.

In social life, as much as it was possible with the autocracy, democratic principles were manifested, which was stimulated by the governor. Neither skin color nor social status were impediment to visit any public places and institutions. A lot of attention was paid to the education of children inculcated attentive attitude and respect for the woman (Scott and Zeuske, 2002). The racial problem at all the difficulty of its solving, especially in connection with a match race and class contradictions, rooted tradition, lost former sharpness. Toussaint was doing his utmost to eliminate the remnants of mistrust and suspicion between members of different races.

In the country the court proceedings has been ordered. For fighting abuse Institute of government commissioners was created, performing functions resembling modern features of prosecutors. Cassational Tribunal was established.

The policy of the new state has provided unprecedented flourishing of the economy of the island. Strict military discipline provided dedication of effort and ensured the smooth organization of work. Liberty contributed to the increase in population, which before were dying out and was replenished mainly due to the importation of slaves from Africa. The liberation from slavery served as a powerful impetus of development of black music and art (Scott and Zeuske, 2002).

According to Carolyn Fick, the revolutionary slaves of Saint Domingue radicalized the notion of natural rights and exposed the tensions and philosophical contradictions in Enlightenment thought in ways that no other historical event of that period could have done. In its particularities, as in its universal qualities, the Haitian revolution turned on its head the principles enshrined in the Rights of Man and, contra distinctively, redefined the meaning of freedom" (Fick, 2013, P.395).

At the same time, "Freedom accompanied by citizenship required that the newly freed, that is, the armed slaves who had been waging warfare since the first outbreak of August 1791, should become responsible citizens" (Fick, 2013, P.396).

During the War of Independence were eliminated many barriers to capitalist development in Latin America (the poll tax, labor service population in favor of the Spanish crown, colonial taxes and customs duties). Class and racial inequality wer canceled, Inquisition was banned, a republican system was established (Craton, 1988).

But the War of Independence has not led to the solution of the agrarian question. In all Independent States large latifundias remained. In the course of the revolution were confiscated estates owned by Spaniards and their supporters. But these lands passed into the

4

hands of a new landed aristocracy, which was formed during the War of Independence from its leaders, especially among the military.

Also estates of the old Creole elite were preserved. Pre-capitalist remnants in the village (mining, peonage, and others) did not disappear.

Politically, the war for independence ended in compromise between still weak bourgeoisie and the new agrarian oligarchy, which made extremely difficult carrying out the bourgeois-democratic reforms in Latin American countries. Some bourgeois reforms were carried out only in the second half of the 19th century as a result of the long and bloody civil war. Therefore, the war for independence in Latin America can be regarded as unfinished bourgeois revolution.

Thus, for the Latin American countries fatal consequences had the fact that they achieved independence in an environment where in developed European countries already under way was the industrial revolution. The backwardness of Latin American countries, the incompleteness of the bourgeois-democratic reforms predetermined their economic and financial dependence on foreign capital.

Specificity of economic and political development in Latin America has been driven by a belated compared with Europe taking the path of bourgeois progress. The giant gap baselines of the Old and New Worlds, due to objective historical reasons, predefined the inclusion of Latin American countries into a single world economic complex, first by colonization and by then unequal relations of dependence on the advanced centers of world capitalism. The basis of the region's rapid transition to capitalism has been its initiation to the world capitalist market as a peripheral agrarian and raw material item.

A characteristic feature of bourgeois development in such circumstances was the fact that here the new social, economic and political structures did not just come to replace the old, but oppressed them, integrated them into their orbit. The ability to integrate the components of the old structures to the new facilitates and accelerates the initiation of these countries to the bourgeois progress, made them amenable to perception of coming from outside the new, advanced forms. The same can be said with regard to culture, social psychology and ideology. As a result, during only four centuries - from the early 16th century to the early 20th century - Latin America achieved the historic leap from the Stone Age primitive communal system and the early civilizations of the ancient Eastern type to the stage of industrial capitalism, for what Europe needed millennium.

The flip side of these processes was the extraordinary vitality of the integrated elements of the old, traditional structures in the new. This, along with the acceleration of bourgeois progress, led to the dominance of its conservative options, rooted multiculturalism, when the formation and development of the capitalist mode of production combined with the conservation of components of pre-capitalist ways of life, with the presence of small-scale, patriarchal economy and even the primitive society of Indian tribes (in areas undeveloped by "civilization"). This increased the contradictions of social development (Schmidt-Nowara, 2000).

Great influence on the socio-political and cultural development of countries in the region have had a particular form of Latin American nations that were the product of an intermingling of the Indian population with European newcomers and natives of Africa. Nation evolved from diverse racial and ethnic components of respectively socio-economic and territorial and state community. In some cases these processes in the beginning of 20th century had not yet ended. For example, single nation has not been developed from Indian and Creole population in Peru. Immigration from Europe continued. Immigrants from Europe and their descendants accounted for the vast majority of residents of Argentina, Uruguay and Costa Rica, more than half of Brazilians and Cubans. Among the descendants of the Europeans dominated the persons of Spanish (in Brazil - Portuguese) origin. There were many immigrants arrived in the late 19th - early 20th century from Italy (Argentina, Uruguay), from the Slavic countries and Germany (Argentina, Brazil, Chile). Indians and mestizos were the main population of Paraguay, Guatemala, Andean highlands - Bolivia, Peru and Ecuador. Mestizo - mixed descendants of Europeans and Indians - prevailed in Mexico, Chile, Venezuela and Colombia, most of the Central American republics. Negros and mulattos became the main population of the colonial possessions of Great Britain and France in the Caribbean, as well as Haiti and the Dominican Republic. From blacks and mulattos more than a third of the population of Brazil and Cuba was comprised. In the Caribbean, there were people from India (British Guiana, Suriname, Trinidad and Tobago) and Chinese (Schmidt-Nowara, 2000).

The interaction of different traditions, cultures, customs, psychological composition - Indian, Negro, European (mainly the southern branch of the Iberian-Roman European civilization) - made it a kind of ethno-cultural fusion. Hispanics were distinguished by characteristic to many southern peoples temperament, a tendency to a bright, emotional manifestations of life. This is reflected in the socio-political struggle bore stormy, especially in a wide range of deep social and economic contradictions and social instability, the presence of mass devastation, disorganized, disadvantaged population, ready for rebellion and revolutionary outbreaks, the despair and passivity or rushing for reformist or conservative-reactionary figures.

Unstable, "the troubled' state of the Hispanic community, lack of "political culture", marked ethnic and psychological traits gave rise to political instability, violence accounts for more weight ("violensiya") in political life. This was evident in the frequent revolts, coups and counter coup, assassinations of political and public figures, mass repression and dictatorships, guerrilla and civil wars, uprisings and revolutions in the rebel-anarchist tendencies in the popular movements, including the actions of the workers and peasants.

Authoritarian, dictatorial regimes dominated. Constitutional, democratic forms of political life, party-political structures were underdeveloped, were unstable and deformed or simply absent.

A characteristic feature of the social and political life of the Latin American republics was the persistence of patriarchal and paternalistic, caudillo (from the word "caudillos" - the leader) traditions, clan, formed in the era of colonialism, provincial isolation and civil war 19th century. The starting point is the prevalence of "vertical" social ties between the "master", "chuck", "leader" and his subordinate mass, or "clientele" of the "horizontal" class and social relations. The essence of such "vertical" relations – is in the cohesion of a group of people around a strong, influential person in the hope to break out after the top of the personality rights of its nearest support in the competition with other similar "clans". This way in ordinary, everyday life seemed more approachable and real. And, accordingly, in the political struggle in the popular movement of the masses united not so much about specific political and ideological platforms, how about leaders who in their eyes are brought bright, strong-willed, "charismatic" personalities, capable to win over, to win both the authorities and provide further on top of the aspirations of their followers. At the forefront the personal qualities of the leader and his ability to capture the psychology, the mood of the masses, the "mob", to appear in front of them "his" close to them, acting not so much on common sense than on feelings and emotions, the subconscious. "Strong personality" asserted his authority in authoritarian political movements, parties in the country, often relying on its own armed force and political clientele. They claim to be the supreme "leaders", the "fathers" of the nation, the people. The masses of illiterate or semi-literate population, particularly outside the major economic and cultural centers, they could not even make up their own "civil society" and social base for participatory democracy. Most Latin American countries were republics rather in name only. In fact, often the republican and constitutional facade covered authoritarian and dictatorial regimes or narrow elitist "democracy" to the exclusion of the general population from real participation in political life.

Also should be mentioned that in the 19th and 20th centuries almost every caudillo was genuinely convinced that people are not ready for democracy on the European model, and considered dictatorial form of government as a necessary and logical stage of political evolution, the natural course of things. The arguments do not differ by novelty and originality: you must first ensure public order and economic progress, and then we can talk about the creation of democratic institutions, protection of rights and freedoms of the individual, etc. Such sentiments are deeply rooted in Latin American political discourse and in a number of states in the region successfully survived to this day, reflecting the remarkable vitality of the theory and practice caudillo.

The main objective of social and political forces in Latin America in 21 century is the search for and implementation of development options that respond to local conditions and would allow countries in the region to combine the renewal of the economy with the interests of the majority of society, with the decision of the most pressing social and political problems, combining integration into the world Community retaining its original identity, its own civilization foundations typical for the peoples of Latin America.

Works Cited

Craton, Michael. Continuity Not Change: The Incidence of Unrest Among Ex-Slaves in the British West Indies, 1838-1876. Slavery & Abolition: *A Journal of Slave and Post-Slave Studies*, 1988, Volume 9, Issue 2, 144-170. Print.

Fick, Carolyn E. The Haitian revolution and the limits of freedom: defining citizenship in the revolutionary era. *Social History*, 2007, 32:4, 394-414. Print.

Scott, Rebecca J. and Zeuske, Michael. Property in Writing, Property on the Ground: Pigs, Horses, Land, and Citizenship in the Aftermath of Slavery, Cuba, 1880-1909. *Comparative Studies in Society and History*, 2002, Vol. 44, No. 4, 669-699. Print.

Schmidt-Nowara, Christopher. The end of slavery and the end of empire: Slave emancipation in Cuba and Puerto Rico, Slavery & Abolition: *A Journal of Slave and Post-Slave Studies*, 2000, 21:2, 188-207. Print.